Fluid Power
Educational
Series

Pipes, Tubes, and Hoses in Hydraulic Systems
(In the English Units)

Joji Parambath

Pipes, Tubes, and Hoses in Hydraulic Systems
(In the English units)

Copyright © 2020 Joji Parambath

All rights reserved

ISBN: 9798653900525

https://jojibooks.com

Disclaimer of Liability

The contents of this book have been checked for accuracy. Since deviations cannot be precluded entirely, we cannot guarantee full agreement. Only qualified personnel should be allowed to install and work hydraulic equipment. Qualified persons are defined as persons who are authorized to commission, to ground, and tag circuits, equipment, and systems following established safety practices and standards.

Table of Contents

PREFACE

Fluid conductors interconnect components of a hydraulic system for the safe and leak-free transmission of high-pressure hydraulic fluid throughout the system. As hydraulic systems are getting more and more complicated with their operation under increased temperatures and in limited spaces, not only the fluid conductors must put up with these adverse conditions, but also handle the high working pressures, peak surge pressures, and peak flow rates. A vast number of hydraulic applications, demands numerous types of conductors to satisfy the varying working requirements and conditions.

This book presents the necessary information about the constructional features, performance specifications, and other details of pipes, tubing, and hoses and their fittings. Next, the topics are logically arranged for a simple to the complex level progression of the subject matter. The book uses the English system of units.

Many other fluid power topics are given in other textbooks under the fluid power educational series by the same author. A list of all the textbooks is given at the end of the book (Page No. 59). Also, please see the details at https://jojibooks.com

Enjoy reading the book.
Your feedback is most welcome.

JOJI Parambath

Chapter 1 | Fluid Conductors – Introduction and Terms & Definitions

Introduction

In a conventional hydraulic system, various components of the system are assembled through a conductor system. That means the conductor system is a network of conductors that connects to the system components through fittings for the effective delivery of the fluid through the system. Pipes, tubing, and hoses are the three basic types of fluid conductors used in hydraulic systems.

A conductor is a pressure-tight vessel used to convey a sufficient quantity of pressurized fluid through it in a leak-free manner. It must have smooth interiors to reduce the friction and flow turbulence during the fluid flow, and sufficient wall thickness to withstand the high operating and shock pressures developed in the system. Further, it must be capable of withstanding the high system and ambient temperatures. It must also be compatible with the type of fluid used.

The critical considerations for the selection of fluid conductors include their construction, sizing, installation, routing, and applicable standards.

Terms and Definitions, Fluid Conductor

The fluid power industry uses a multitude of conductor-related terms to specify the performance levels of fluid power systems. For example, a fluid conductor is specified by its diametrical size and wall thickness. The following sections present a brief account of some commonly used terms and definitions of fluid conductors.

Diametrical Size

The diametrical size of a conductor is specified by its inside diameter, outside diameter, or nominal size (Figure 1.1).

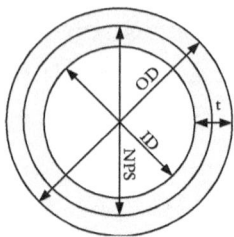

Figure 1.1 | Size specifications of a fluid conductor

Inside Diameter, D_i
It is the smallest cross-sectional diameter of a conductor.

Outside Diameter, D_o
It is the largest cross-sectional diameter of a conductor.

Nominal Size
In ANSI and SAE standards, for example, the size of a pipe is specified in terms of nominal pipe size (NPS), and in the SI system, it is specified in terms of Nominal Diameter (DN). A detailed explanation of this topic is given in chapter 2 under the heading 'Nominal Pipe Size and Diameter Nominal'.

Wall Thickness, t
The wall thickness of a pipe or tubing decides the maximum pressure that can be subjected to it. It is expressed in terms of metric units or a schedule number as in ANSI and SAE standards. It is given by:

$$\text{Wall thickness, } t = (D_o - D_i) / 2$$

Schedule Number
The schedule numbers vary from 5 through 160 in a graded manner. Altogether, there are eleven different schedule numbers. They are: 5, 10, 20, 30, 40, 60, 80, 100, 120, 140, and 160.

Larger schedule number for a given pipe size points to more substantial wall thickness. For a particular size of the pipe, the outer diameter stays the same, but the inside diameter becomes smaller as its schedule number increases.

As an example, Table 1.1 gives the wall thicknesses corresponding to schedule numbers 40, 80 and 160, for a pipe of nominal size ½. It can be observed that the wall thickness increases as the schedule number increases. The topic is further elaborated in chapter 2 under the heading 'Wall Thickness – Schedule Numbers'.

Table 1.1 | Wall thicknesses of pipes

Nominal Pipe Size		Outside Diameter	Wall thickness		
			Schedule 40	Schedule 80	Schedule 160
		inch	inch	inch	inch
½	0.500	0.840	0.109	0.147	0.188

Example 1.1
Determine the inside diameter of a pipe with an OD of 1.9 inch and a wall thickness of 0.145 inch.

Solution

Pipe OD	= 1.9 inch
Wall thickness	= 0.145 inch
Inside diameter of the pipe	= 1.9 – 2 x 0.145
	= 1.61 inch

Hoop Stress
Hoop stress in a pipe or tubing is the circumferential stress acting on the wall of the conductor and trying to split it. It is the maximum pressure that the material of the conductor is capable of withstanding before pulling apart. Consider a thin-walled conductor of the inside diameter (D_i), wall thickness, (t, where t < $0.1 \times D_i$), and length (L), as shown in Figure 1.2. The conductor is subjected to operating pressure, P.

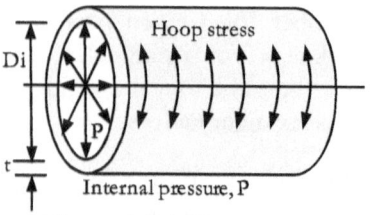

Figure 1.2 | Hoop stress

The operating pressure, acting normal to the inside surface of the pipe, induces a circumferential force in the conductor that tends to split the conductor into two halves. The normal surface area can be taken as the projected area (D_i x L) of one half of the pipe. The circumferential force (or burst force) is given by:

$$\text{Circumferential force} = P \times D_i \times L$$

The tensile force which is trying to resist the splitting of the conductor acts on the cross-sectional area (tL) of each wall. Therefore, the resistive force is given by:

$$\text{Resistive force} = 2tL \times \text{Hoop stress.}$$

Equating the circumferential force to the resistive force, we get,

$$P \times D_i \times L = 2tL \times \text{Hoop stress}$$

Therefore,

$$\text{Hoop stress} = P \times D_i \ / \ 2t$$

The conductor must have sufficient tensile strength to prevent its bursting due to the excessive hoop stress.

Tensile strength: It is the ability of a material to withstand a pulling (tensile) force

Yield Strength: It is the stress a material can withstand without permanent deformation

Example 1.2
Determine the hoop stress developed in a pipe of outside diameter 2.375 inch and a wall thickness of 0.218 inch when the pipe is subjected to a pressure of 1000 psi.

Solution
Outside diameter of the pipe, Di = 2.375 inch
Wall thickness of the pipe, t   = 0.218 inch
Pressure, P   = 1000 psi
Hoop stress developed   = P x D_i / 2t
                                                      = 1000 x 2.375 / 2 x 0.218
                                                      = 5447 psi

Burst Pressure
It is the internal pressure inside a fluid conductor that causes it to burst or rupture (Figure 1.3). The conductor bursts when the hoop stress exerted on the conductor exceeds the tensile strength (S) of the conductor material. Barlow's formula is commonly used to predict the burst pressures in ductile thin wall tubes (t < 0.1 x D_i).

$$\text{Burst pressure (BP)} = 2tS / D_i$$

For thick-walled pipes, the tensile stress across the wall thickness is not uniform. Therefore, the following formula must be used to take into account the non-uniform tensile stress.

$$\text{Burst pressure (BP)} = 2tS / (D_i + 1.2t)$$

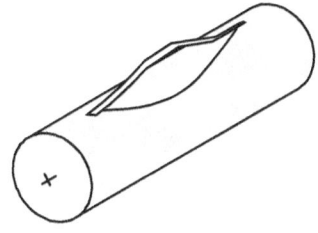

Figure 1.3 | Burst tested tube

Working Pressure

The working pressure of a conductor is the safe pressure to which it can be subjected. It is calculated by dividing the burst pressure of the conductor with a safety factor.

$$\text{Working pressure (WP)} = \frac{\text{Burst pressure (BP)}}{\text{Safety factor (SF)}}$$

- A safety factor of 4:1 is used for hydraulic applications where shock and mechanical strain are not considerable.
- A safety factor of 6:1 should be used where considerable shock and mechanical strain are expected.
- A safety factor of 8:1 should be used where severe hydraulic shock and mechanical strain are expected.

Design Pressure

It is the pressure to which each component of a piping system is designed. It is not to be less than the actual pressure at the most severe condition of pressure and temperature expected during the service of the piping system.

Maximum Allowable Working Pressure

It is the maximum pressure of a piping system, determined by the weakest component of a piping system. It is not to exceed its design pressure.

Minimum Bend Radius

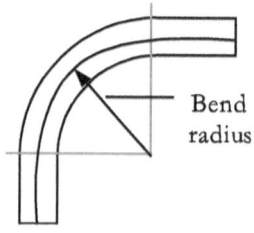

Figure 1.4 | Minimum bend radius

6

It is the smallest radius of the curved section of a conductor (tube or hose) beyond which it should not be bent without flattening, kinking, or wrinkling (Figure 1.4). The bending of the conductor beyond the limit causes severe backpressure and damages the conductor internally leading to its premature failure.

Design Temperature
It is the maximum temperature at which each piping component is designed to operate. It may be taken as the maximum fluid temperature.

Example 1.3
Determine the maximum pressure rating of a thick-wall pipe of OD of 1.315 inch with a wall thickness of 0.25 inch. The pipe material is carbon steel with a tensile strength of 50000 psi. Assume a safety factor of 4.

Solution:

Pipe OD, D_o	= 1.315 inch
Wall thickness, t	= 0.25 inch
Tensile strength, S	= 50000 psi
Safety factor, SF	= 4

Pipe ID, D_i	= D_o – 2 x t
	= 1.315 – 2 x 0.25
	= 0.815 inch

Burst pressure	= 2ts / (D_i + 1.2t)
	= 2 x0.25 x50000/(0.815 + 1.2 x 0.25)
	=22422 psi

Pressure rating	= Burst pressure/Safety factor
	= 22321 / 4 = 5580 psi

Flow Rate
Volumetric flow rate is a measure of the volume of oil passing a cross-section per unit of time. It is usually measured in gpm.

Flow Velocity

The velocity of fluid flow is the average speed at which its particles move past a given cross-section. It is a design parameter that decides the internal diameter of the pipelines and the type of fluid flow.

Flow Rate Vs Velocity of Flow

Figure 2.8 illustrates the fluid flow through pipe sections having different cross-sectional areas. The flow rate (Q) equals the pipe area (A) multiplied by the velocity (v) of the fluid flow. That is,

$$Q \ (ft^3/s) = A(ft^2) \ x \ v(ft/s)$$

The relation for the fluid velocity in commonly used units are given by:

$$v(ft/s) = \frac{0.3208 \ x \ Q \ (gpm)}{A \ (in^2)}$$

Example 1.4

The fluid is flowing through a pipe of an internal diameter of 1" size at the rate of 20 gpm. Determine the fluid velocity.

Solution

Pipe size   = 1"

Flow rate   = 20 gpm

Area of cross-section $= \pi \ . \ D^2/4$

                               $= 3.14 \ x \ 1 \ x \ 1 \ /4$

                               $= 0.785 \ in^2$

$$v(ft/s) = \frac{0.3208 \ x \ Q \ (gpm)}{A \ (in^2)}$$

$$v(ft/s) = \frac{0.3208 \ x \ 20}{0.785}$$

$$= 8.17 \ ft/s$$

Pipe and Tube Materials

The pipe and tube materials suitable for high-pressure industrial hydraulic service are: (1) Cold-drawn seamless carbon steel and (2) Cold-drawn seamless stainless steel. Stainless steel pipes and tubes are used in applications that require resistance to corrosion, such as in chemical equipment or marine vessels. Hot rolled pipes are not recommended for hydraulic services as they leave behind scales inside and outside.

Steel is an alloy of iron and carbon (<2%). Steels are classified into four basic types according to the amount of carbon and other alloys. They are carbon steel, alloy steel, stainless steel, and tool steel.

- Carbon steel is an iron-carbon alloy. It also contains manganese, silicon, and copper, lead, aluminum, cobalt, chrome, nickel, etc. in varying degrees. Carbon steels are vulnerable to corrosion.

- Alloy steel is prepared by adding steel with various metals such as iron, nickel, aluminum, copper, etc. The strength and properties of alloy steel depend on the amount of the elements present in the alloy steel.

- Stainless steel has low carbon content but contains chromium alloy, and nickel or molybdenum. It is strong and corrosion-resistant, and it can withstand high temperatures.

- Tool steels are hard and used to make metal tools and mold-making tools.

Steel specifications are given by different organizations like ISO, AISI, SAE, ASTM, European Standard (EN), German standard (DIN), Japanese Industrial Standards (JIS), etc. in their unique ways.

Properties and Standards of Typical Steel Materials

Table 1.2 | Properties and standards of typical steel materials

Pipe material	Properties	Standards
Cold-drawn seamless carbon steel	High-pressure capability, precise dimensions/shape, clean inside surface with no scale, excellent scaling surface after roll flaring	DIN EN 10305-4 E 355N (St. 52.4 NBK) E 235N (St. 37.4 NBK)
Cold-drawn seamless stainless steel	High-pressure capability, precise dimensions/shape, excellent scaling surface after roll flaring	DIN EN 10216-5 ASTM A269/A213 ASTM A312

Tensile Strengths and Yield Strengths of Steels

Table 1.3 | Tensile strengths and yield strengths of steels

Steel type	Tensile strength (N/mm^2)	Yield strength $(N/mm^2$ min$)$
E235N tubes (St 37.4)	340	235
E355N tubes (St 52.4)	490	355
AISI 316L metric size tubes	485	170
TP 316L schedule size pipes	485	170
DIN 2391 St. 45	570	255
DIN 2391 St. 52	630	355

E – Steel for machine parts
235 – Minimum yield strength in N/mm2

Chapter 2 | Hydraulic Pipes and Fittings

Pipes are rigid conductors with relatively larger wall thickness used to contain and convey hydraulic fluids. They are highly resistant to bending. It is difficult to shape rigid pipes into the desired configuration. Remember, configuring a piping system is more labor intensive. Many fittings, such as elbows, tees, etc. are needed to be used while routing a piping system. They are liable to transmit shock and vibration from one component to another. All piping segments should be secured with clamps, and preferably with damped one, to absorb the shock and vibration and prevent their propagation.

Constructional details, Pipes
Pipes have a comparatively larger wall thickness, but they are much cheaper than tubes and hoses. Hence, they are generally employed in applications where there is no restriction on the size of the conductors, and cheaper conductor systems are preferred. It is recommended to use pipes with higher tensile strength, as far as possible, because higher tensile strength means higher permissible working pressures and reduced wall thickness, leading to reduced overall weight in the pipe and necessary supporting structures. Pipes are usually not used in mobile hydraulic systems.

Cold-drawn seamless carbon steel pipes and austenitic stainless steel pipes are used in hydraulic systems due to their strength. With these types of materials, the pipes can be shaped and sized precisely. They maintain good cleanliness without any scale. In general, carbon steel is used for pipes employed in indoor hydraulic applications. Stainless steel pipes are used in applications that require resistance to corrosion, such as chemical equipment or marine vessels. A galvanized pipe is not recommended for use in hydraulic systems.

Advantages and Disadvantages, Pipes

Pipes have many advantages and disadvantages. Some of them are listed below:

- A conductor system with steel pipes is the least expensive way to assemble a hydraulic system with low to medium pressure ratings
- As pipes are made of inflexible material and have a large wall thickness, they are difficult to form into the desired configuration and install
- They cannot withstand high surge pressures

Thermal Expansion of Pipes

There can be significant variations in temperature in hydraulic systems, especially in marine and offshore applications. Under certain conditions, the temperature can vary widely, for example, from −40°C during winter to +40°C during summer. The variations in temperature results in the thermal expansion of the pipes. Remember, a pipe length can vary almost 1 mm per 1-metre length of the pipe with a temperature difference of 80°C.

Basic Requirements, Pipes

The basic requirements of pipes for hydraulic service are enumerated below:

- Pipes must have sufficient cross-sectional areas to satisfy the flow rate requirements without producing excessive pressure drops
- They must be strong enough to withstand the working pressure, shock pressures, and vibration
- They should have smooth interiors to reduce the friction and flow turbulence
- They must be compatible with the type of fluid used
- They must withstand high operating temperatures
- They must be supported by damped mountings to absorb both shock and vibration

Size Specifications

A pipe should have sufficient ID and smooth inside surface to reduce frictional forces. The wall thickness of the pipe decides its pressure rating. The optimum pipe size should be based on minimizing the sum of energy cost and piping cost.

Pipe sizes are standardized to reduce the numbers of pipe sizes. In current practice, pipe size is defined with two sets of numbers. They are: 1) Pipe schedule (wall thickness) and 2) Nominal pipe size (NPS) or Nominal diameter (DN). NPS is also referred to as NB (Nominal Bore).

Wall Thickness – Schedule Numbers

Manufacturers offer standard and non-standard sizes. Schedule numbers 40, 80 and 160 are most commonly used for specifying the wall thickness of pipes for hydraulic systems.

- The schedule number 40 conforms to the 'standard' wall thickness intended for low pressures.
- The schedule number 80 conforms to the 'extra heavy' wall thickness intended for high pressures.
- The schedule number 160 conforms to the 'double extra heavy' wall thickness.

The schedule number indicates the approximate value of the expression 1000 x P/S, where P is the service pressure, and S is the allowable stress, both expressed in the same unit.

Nominal Pipe Size (NPS)

It is a size standard established by the American National Standards Institute (ANSI). It is the number that defines the size of the pipe. For a 6 NPS pipe, the 6" is the nominal size. Nominal sizes of pipes do not always correspond to its inside diameter or outside diameter. Hydraulic pipes come in nominal sizes from 1/8" to 42". Each size is available in a variety of wall thicknesses. Table 2.1 presents the minimum wall thickness corresponding to different sizes for steel pipes.

13

Table 2.1 | Standard nominal pipe sizes and dimensions

Nominal Pipe Size	Outside Diameter	Wall thickness		
		Schedule 40	Schedule 80	Schedule 160
--	inch	inch	inch	inch	inch
⅛	0.125	0.405	0.068	0.095	--
¼	0.250	0.540	0.088	0.119	--
⅜	0.375	0.675	0.091	0.126	--
½	0.500	0.840	0.109	0.147	0.188
¾	0.750	1.050	0.113	0.154	0.219
1	1.000	1.315	0.133	0.179	0.250
1¼	1.250	1.660	0.140	0.191	0.250
1½	1.500	1.900	0.145	0.200	0.281
2	2.000	2.375	0.154	0.218	0.344
2½	2.500	2.875	0.203	0.276	0.375
3	3.000	3.500	0.216	0.300	0.438
3½	3.500	4.000	0.226	0.318	--
4	4.000	4.500	0.237	0.337	0.531
5	5.000	5.563	0.258	0.375	0.625
6	6.000	6.625	0.280	0.432	0.719
8	8.000	8.625	0.322	0.500	0.906
10	10.00	10.75	0.365	0.594	1.125
12	12.00	12.75	0.406	0.688	1.312
14	14.00	14.00	0.438	0.750	1.406
16	16.00	16.00	0.500	0.844	1.594
18	18.00	18.00	0.562	0.938	1.781

As can be observed from Table 2.1, for a pipe with a nominal size in the range from 1/8" to 12", the internal diameter (ID) of the pipe is not the same as the nominal pipe size but deviates from it depending on its schedule number. For a pipe with a nominal pipe size of 14" or above, the pipe OD corresponds to the nominal pipe size. It may be noted that any increase in the wall thickness decreases the inside diameter of the pipe.

Pipe Schedule for Stainless Steel Pipe

The cost of stainless steel pipe is much higher than that of the carbon steel pipe. Due to corrosion resistance properties of stainless steel, the advancement of high alloy stainless steel and fusion welding, a thinner steel pipe can work satisfactorily without fear of early failure.

To reduce the cost of the pipe material, the American Society for Mechanical Engineers (ASME) has introduced different schedule numbers for stainless steel pipe and fittings.

Under ASME B36.19 Schedule number with 'S' suffix is introduced for stainless steel pipes. Schedule numbers 5S, 10S, 40S, 80S are used for the wall thickness of stainless steel pipe as per ASME B36.19.

Example 2.1

Refer to Table 2.1. Calculate the IDs of schedule 40 pipes corresponding to the following nominal pipe sizes: (1) 2" and (2) 14."

Solution

1.

NPS	= 2"
OD	= 2.375"
Wall thickness (t)	= 0.154
ID	= OD – 2 x t
	= 2.375 – 2 x 0.154
	= 2.067"

2.

NPS	= 14"
OD	= 14"
Wall thickness (t)	= 0.438
ID	= OD – 2 x t
	= 14 – 2 x 0.438
	= 13.124"

Diameter Nominal (DN)

It is the metric equivalent of NPS used to specify pipe sizes. Note, the metric designations conform to ISO as well as European standards.

Table 2.2 gives the Standard metric pipe sizes and dimensions.

Table 2.2 | Pipe sizes in Diameter Nominal

Nominal size	Outside Dia, mm	Wall thickness, mm				
		A	B	C	D	E
6	10.2	1.6				
8	13.5	1.8				
10	17.2	1.8				
15	21.3	2.0	2.8			
20	26.9	2.0	2.8			
25	33.7	2.0	3.2	4.2	6.3	6.3
32	42.4	2.3	3.5	4.2	6.3	6.3
40	48.3	2.3	3.5	4.2	6.3	6.3
50	60.3	2.3	3.8	4.2	6.3	6.3
65	76.1	2.6	4.2	4.2	6.3	7.0
80	88.9	2.9	4.2	4.2	7.1	7.6
90	101.6	2.9	4.5	4.5	7.1	8.1
100	114.3	3.2	4.5	4.5	8.0	8.6
125	139.7	3.6	4.5	4.5	8.0	9.5
150	168.3	4.0	4.5	4.5	8.8	11.0
200	219.1	4.5	5.8	5.8	8.8	12.5
450	457.0	6.3	6.3	6.3	8.8	12.5

Pipe Standards
Steel pipes according to DIN Standards
- DIN EN 10305-4. E 355N (St. 52.4 NBK) or E 235N (St. 37.4 NBK)
- DIN EN 10216-5

1. Seamless cold-drawn precision steel pipes E235 (St 37.4) and E355N(St 52.4) according to EN 10305-4
- E235N (St 37.4) – NBK – Normalized, phosphated and oiled inside and outside
- E235N (St 37.4) – NBK/ZN – Normalized, electric zinc plated with Cr-VI-free passivation
- E355N (St 52.4) – NBK – Normalized, phosphated and oiled inside and outside
- E355N (St 52.4) – NBK/ZN – Normalized, electric zinc plated with Cr-VI-free passivation

2. Austenitic stainless steel pipes, seamless, cold-drawn 316L (1.4404) according to ASTM-standards
- AISI 316L – metric sizes fully annealed, scale-free
- TP 316L – schedule sizes fully annealed, scale-free

Pipe Fittings
Pipe connections are coupled through welded joints, flanged joints, or threaded joints. The type of jointing technology is selected based on the working pressure, pipe size, pipe material, fitting standards, and other conditions such as possible pressure shocks in the system, nature of the environment, etc. Fittings are available as per various standards including NPTF, JIC etc.

Sections of pipes can be joined together using pipe fittings, such as sleeves, elbows, tees, bends, etc. (Figure 2.1). The welded connections are more commonly used in systems involving severe mechanical load, high pressure, vibration, or high temperature. They can also be employed where leaks cannot be tolerated.

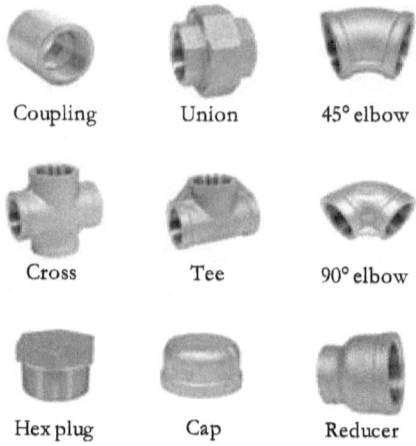

Figure 2.1 | An assortment of pipe fittings

The welded connections can be butt welded joints, socket welded joints, or slip-on welded sleeve joints. The socket weld is used for pipes of size less than 2 inches, whereas the butt weld is used for larger pipes.

However, in hydraulic piping systems with high-quality requirements, it is recommended to use non-welded connection technologies (fittings, flanges, etc.) for all pipe sizes due to their reliability and inherent cleanliness. The threaded connections are most common and are used in applications with pressures up to 2500 psi.

Welded Pipe Joints
Butt-welded joints with complete penetration at the root can be used for joining sections of pipe. Socket-welded joints or slip-on welded sleeve joints, as shown in Figure 2.2(a), may be used for piping with a nominal diameter up to 3 inches, except for toxic or corrosive services. In the case of the slip-on welded sleeve joints, the pipe sections are inserted into the sleeve and welded, as shown in Figure 2.2(b). The fillet weld leg size is to be at least 1.1 times the nominal wall thickness of the pipe.

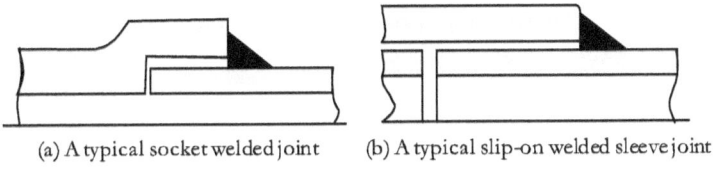

(a) A typical socket welded joint   (b) A typical slip-on welded sleeve joint

Figure 2.2 | Welded pipe joints

Flanged Joints

Pipe sections can also be joined by using flanges. Flanges are to be attached to the pipes by welding or tapered threads. (Figure 2.3)

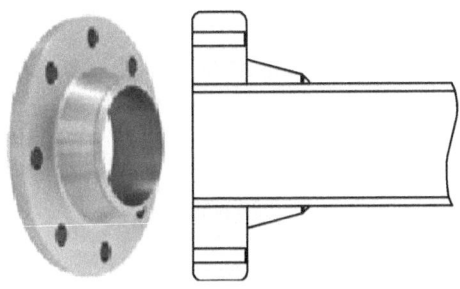

Figure 2.3 | Flanged joints

Thread Joints for Pipes

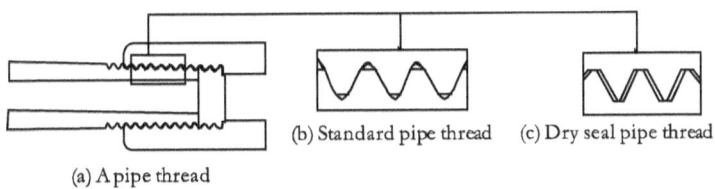

(b) Standard pipe thread   (c) Dry seal pipe thread

(a) A pipe thread

Figure 2.4 | Pipe threads

Thread joints are used for hydraulic service to produce a leak-proof metal-to-metal seal. They are either tapered or straight. The pipe threads are made pressure-tight by sealing on the threads. Pipe threads used in hydraulic piping can be divided into two types: (1) Standard pipe threads and (2) Dry-seal pipe thread. Figure 2.4 shows these types of pipe threads.

Standard Pipe thread

The standard pipe thread has tapered threads that produce a metal-to-metal seal, as shown in Figure 2.4(b). The taper is 1/16 of an inch. The connection is made pressure-tight by the sealing on the threads. This type of threads leaves a spiral clearance as the pipes are tightened. Pipe threads require sealant like Teflon tape or joint compounds to fill any voids between the two threads and make the joint leak-proof. However, these threads often develop leaks gradually, which are difficult to repair in the field.

Tapered pipe threads are used for normal hydraulic services (except toxic or corrosive) for pressures up to 1450 psi and temperature up to 923°F. They can be used for hydraulic services for connection to equipment only, such as pumps, valves, cylinders, accumulators, gauges, and hoses.

Dry-seal pipe thread

When larger pipes are used, dry-seal pipe threads are most appropriate. In this type of threads, pressure-tight joints are not made on the threads. Both threads are parallel, and the sealing is affected by the compression of a soft material onto the external thread, as shown in Figure 2.4(c). When tightened, the dry-seal thread eliminates the spiral clearance. This type of thread form tends to minimize thread leaks.

Straight-thread with 'O' Ring

Straight-thread with 'O' ring type fittings, as shown in Figure 2.5, may be used for connections to equipment, without pressure and service limitations. This type may not be used for joining sections of pipe.

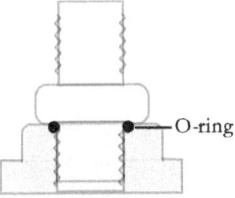

Figure 2.5 | Straight-thread with 'O' ring

Pipe Supports

When designing supports for piping, the following factors should be taken into account:

- The pipes shall not be supported from other pipes nor should the pipes be utilized to support other components

- The transfer of vibration from other equipment and machinery should be avoided to the extent possible

- Thermal expansions shall be taken into account when designing the supports

- A bend should be supported as close to the bend as possible (whenever possible on both sides of the bend)

- The support should be located as close to the end of the pipe as possible when connecting to hose

Tests

All piping systems are to be tested under the working conditions after installation and checked for leakage. Pipes and integral fittings after completion of fabrication may be hydrostatically tested at a test pressure of 1.5 times the design pressure. Small pipes of less than 0.6" outside diameter are usually exempted from the hydrostatic test.

Chapter 3 | Hydraulic Tubing

Tubing is the most widely used type of conductor in hydraulic systems. Tubing is generally a small-diameter thin wall pipe. It can be bent into almost any shape, thus reducing the number of tube fittings while configuring a conductor system. Note, a conductor system with tubing is easier to handle. However, it is usually more expensive than the piping system due to its tighter manufacturing tolerances.

Tubing Construction

Seamless tubing, as shown in Figure 3.1, is formed by cold drawing a billet over a piercing rod. A welded tube is made by forming a piece of cold-rolled steel into a tube and then joining along its longitudinal seam by a material-fusion process. Tubing may have to be pre-formed before installation. Although tubing might make for neater appearance if installed correctly, it requires a lot of planning and skill to bend tubing correctly.

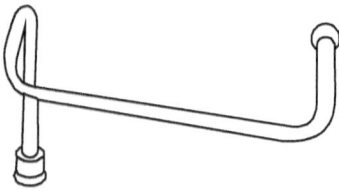

Figure 3.1 | A tubing section

Specifications of Tubing

Essential specifications for hydraulic tubing include its size, pressure ratings, and minimum bend radius. Tubing size is always specified by its outside diameter (OD). Available sizes in English units include 1/16-in increments from 1/8-in outside diameter to 3/8-in. From 3/8-in to 1-in the increments are 1/8 in. For sizes beyond 1-in, the increments are ¼-in. A dash number represents the outside diameter (OD) of tubing expressed in terms of sixteenths of an inch.

Size and Pressure Chart for Carbon Steel hydraulic tubing (Inch Sizes)

Table 3.1

Tube OD (inch)	Wall thickness (inch)	Max. Working Pressure @ 6:1 SF (psi)	Burst Pressure (psi)
3/16	0.035	3422	20533
1/4	0.035	2567	15400
1/4	0.049	3593	21560
5/16	0.049	2875	17248
5/16	0.065	3813	22880
3/8	0.049	2396	14373
3/8	0.065	3178	19067
1/2	0.049	1797	10780
1/2	0.065	2383	14300
1/2	0.083	3043	18260
5/8	0.065	1907	11440
5/8	0.095	2787	16720
3/4	0.049	1198	7187
3/4	0.065	1589	9533
3/4	0.095	2322	13933
3/4	0.109	2664	15987
1	0.065	1192	7150
1	0.095	1742	10450
1	0.120	2200	13200
1-1/4	0.095	1393	8360
1-1/4	0.120	1760	10560

Pressure Rating, Tubing

Burst pressure is the point at which a tube will rupture at a given pressure value. Working pressure is the value considered safe to operate the system under normal working conditions. The safety factor is the ratio between working pressure and burst pressure. As a general rule, a good safety factor is at least 4 : 1.

Tubing Material

Tubing is constructed of dead soft cold-drawn carbon steel has become the accepted standard for use in hydraulics, because they have the mechanical properties required to withstand the high pressures (tensile strength of 55,000 psi). If greater strength is required, the tube can be AISI 4130 steel, which has a tensile strength of 75,000 psi. Copper and brass tubing are not recommended for hydraulic plumbing as these materials harden under vibration and hydraulic shock, and are likely to be subjected to adverse chemical reactions with specific fluid contaminants.

Example 3.1
Determine the inside diameter of tubing section with an OD of 0.5 inch and a wall thickness of 0.049 inch.

Solution:

Pipe OD, Do = 0.5 inch
Wall thickness, t = 0.049 inch
Inside diameter of the pipe = Do – 2t
$$= 0.5 – 2 \times 0.049$$
$$= 0.402 \text{ inch}$$

Example 3.2
Determine the hoop stress developed in a tubing of outside diameter ½ inch and a wall thickness of 0.083 inch when the tubing is subjected to a pressure of 3000 psi.

Solution

Outside diameter of the tubing, Di = 0.5 inch
Wall thickness of the tubing, t = 0.083 inch
Pressure, P = 3000 psi
Internal Diameter = 0.5 – (2x0.083)
$$= 0.334 \text{ inch}$$
Hoop stress developed = $P \times D_i / 2t$
$$= 3000 \times 0.334 / 2 \times 0.083$$
$$= 6036 \text{ psi}$$

Example 3.3
Determine the maximum pressure rating of tubing of OD of ¾ inch with a wall thickness of 0.049 inch. The tube material is carbon steel with a tensile strength of 50000 psi. Assume a safety factor of 6.

Solution:

Tube OD, Do	= 0.75 inch
Wall thickness, t	= 0.049 inch
Tensile strength, S	= 50000 psi
Safety factor, SF	= 6
Tube ID, Di	= Do – 2 x t
	= 0.75 – 2 x 0.049
	= 0.652 inch
Burst pressure	= 2ts / Di
	= 2 x 0.049 x 50000 / 0.652
	=7515 psi

Pressure rating of the tube = Burst Pressure / Safety factor
= 1569 / 6 = 1252 psi

Minimum Bend Radius
It is the smallest radius of the curved section of a fluid conductor beyond which it cannot be bent without flattening, kinking, or wrinkling. Bending of the conductor beyond the limit causes excessive flattening of the tubing in the bend region and damages the conductor internally leading to its premature failure. It also causes severe backpressure.

The bend radius is typically measured along the centerline of the tubing, as shown in Figure 3.2. It is measured as the distance from the center of curvature of the tubing to the centerline of the tubing. A rule of thumb suggests a minimum bend radius of three times the outside diameter. Tube bending is to be following relevant standards.

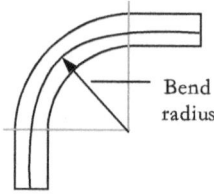

Figure 3.2 | Minimum bend radius of tubes

Tubing Bending Process

Hydraulic tubing can be bent by hand or by using a power bending equipment. Steel tubing can be bent by unique methods, such as roll forming, press forming, mandrel bending, or table forming. In a tube bending process, the pressure is applied to bend the tubing around a correctly-sized die to form the required radius. A tube has a standard bend radius.

Selection of Tube

Proper tube material, type and size for a given application and type of fitting are critical for efficient and trouble-free operation of the associated fluid power system. The selection of proper tubing involves choosing the right tube material and determining the optimum tube size (O.D. and wall thickness). The proper sizing of the tube for various parts of a hydraulic system results in an optimum combination of efficient and cost-effective performance.

Advantages and Disadvantages, Tubing

The main advantage of tubing is that it can be bent into shape and thus requires fewer fittings. Fewer connections generally mean there will be less possibility of leaks. Tubing is also known for its ability to absorb vibration and has a smooth inside finish which is suitable for fluid movement.

Tube Fittings

Since the wall sections of tubing are relatively thin, threading cannot be used to seal the tubing connections. There are varieties of tube fittings available for hydraulic applications. Tubes can be joined quickly and easily with flaring, brazing or couplings. Flared or flareless-type fittings are used for tubing end-connections.

For tube OD from 3/8" to ¾" and pressures up to 3000 psi, flareless end forming connection technology is used. For tube OD from 1" to 1¼", 37° or 45° flaring connection technology is used, generally. The tubing is usually flared to JIC 37° (for >1000 psi) or SAE 45° (for <1000 psi). The flare angle of 45° is used in automotive and refrigeration work, but, not used on hydraulic plumbing.

Flare Fitting, Tubing

It is made up of a nut and a sleeve over the flared tubing, and a body, as shown in Figure 3.3. When forming flares, it is necessary to prepare the tubing - cut square, file smoothly, remove burrs.

The most critical step in making a flare tube fitting is forming the flare without galling, over-thinning, or splitting the end of the tube. The sleeve and nut are pushed smoothly over the tubing end. The sleeve prevents the nut from twisting when the nut is tightened. When the nut is screwed onto the body, it draws the sleeve and the flare against the body, thus forming a seal.

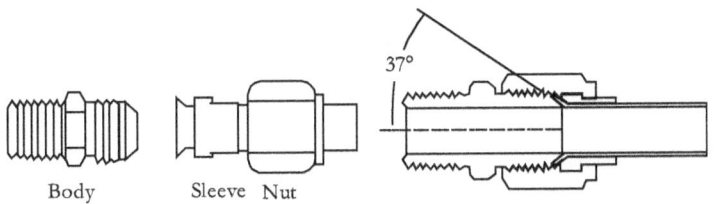

Figure 3.3 | Flare fitting

Compression (flareless) Fitting, Tubing

It consists of a body, ferrule(s) and a nut, as shown in Figure 3.4. First, the ferrules and nut be slipped over the tubing. The tubing is inserted into the body, where it butts up against the shoulder. When the nut is screwed onto the body, the ferrule bites into the skin of the tubing to achieve the holding ability of the connection. This tight connection provides a positive seal. It is used on medium and heavy wall tubing or when tubing cannot be flared.

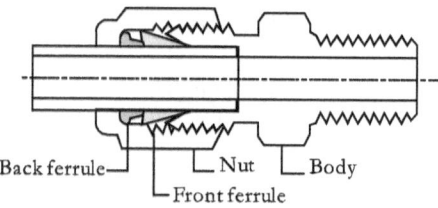

Figure 3.4 | Compression fitting

Selection of Tubes and Fittings

Proper tube material, type and size for a given application and type of fitting are critical for efficient and trouble-free operation of the associated fluid power system. Selection of proper tubing and fittings involves choosing the right tube material and determining the optimum tube size (O.D. and wall thickness). Proper sizing of the tube for various parts of a hydraulic system results in an optimum combination of efficient and cost-effective performance.

Chapter 4 | Hydraulic Hoses

Hoses are the most flexible and versatile type of conductors. They are capable of bending and flexing easily. Hoses can accommodate vibration and pulsations better than tubing. They are selected, when rigid pipes or semi-rigid tubes cannot be used, as in applications with components that move relative to each other or when there are excessive vibration and constant pressure pulsations.

A hose assembly consists of a hose and end fittings that connect directly to adjoining pipe-work or fittings. A hose assembly must have correct end-fitting configurations. It is easy to install a hose assembly along with a well thought out routing layout. Given its superior routing advantage, a hose is generally preferred over metal tubing.

Hose Construction
Hydraulic hoses have three parts: inner tube, reinforcement layer, and protective cover, as shown in Figure 4.1.

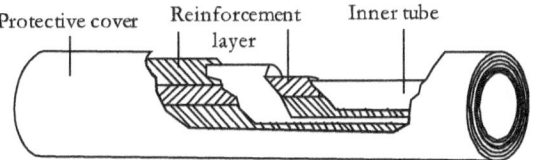

Figure 4.1 | A cut section of a hose

Inner Tube
The inner Tubing is the hose lining, which comes into direct contact with the fluid. Therefore, it must be chemically resistant to the fluid. It must also withstand the effects of extreme variations in fluid temperature. It is made of synthetic rubber, thermoplastics, or PTFE.

Reinforcement Layer

The Reinforcement Layer, as shown in Figure 4.2, provides the strength to withstand the internal pressures and external forces. This layer is made of steel wires, textiles, or synthetic materials. It can also be a combination of wire and textile. Further, the layer determines the working pressure inside the hose.

Figure 4.2 | Different types of hose reinforcement layers

The reinforcement layer is constructed with a single layer or multiple layers of braids or spirals. It can be made with four or six layers to meet severe application demands. This spiral reinforcement is particularly well suited to high-pressure impulse applications. A hose with multiple reinforcements may be provided with an anti-friction layer between them to prevent the steel wires from rubbing against each other. The reinforcement layer of the hose, connected to the suction side of a pump, can also be made with a helical coil to keep the hose from collapsing due to the existence of partial vacuum during the fluid suction process. This type of construction also keeps the hose from collapsing due to tight bending.

Braided reinforcement can be of steel wire or textile or a combination of wire and textile. The reinforcement layer may have single or multiple layers. The type of reinforcement depends on the intended use of the hose.

Protective Layer

The primary purpose of the cover is to protect the tube and reinforcement from abrasion, corrosion, extreme temperatures, UV light, and ozone. The cover can be made from synthetic rubber, fiber braids, or a combination of both depending on the application. Hoses with synthetic rubber covers are generally preferred over textile-braid covers because they are more resistant to abrasion.

Types of Hoses by Operating Pressures

Another way of classifying hydraulic hoses is based on their pressure ratings. The classification is presented below:

- **Low pressure hoses**: They are designed for use in various applications with operating pressures below 300 psi. Their reinforcement is usually textile.

- **Medium-pressure hoses:** They are used for hydraulic applications requiring operating pressures of 300 psi to 3,000 psi. They may be a one-wire braid or multiple wires and/or textile braid construction.

- **High-pressure hoses:** They are frequently found in high-pressure hydraulic applications such as construction equipment requiring operating pressures of 3,000 psi to 6000 psi. These hoses are often called 'two-wire' braid hose because they generally have a reinforcement of two-wire braids of high tensile strength steel.

Suction Hose

A hose connected to suction service is subjected to crushing forces because the atmospheric pressure outside the hose is higher than the internal pressure. They also tend to restrict the flow through it. Therefore, hoses connected to the pump suction line must resist its collapse due to the pressure differential across it. The best way to prevent the hose collapse is to reinforce the hose with a helical wire. The size and spacing of the helical wire reinforcement depend on the size of the hose and the pressure differential.

Very High-pressure Hoses

They are used for off-highway equipment and heavy-duty machinery where extremely heavy pressure surges are encountered. The oil-resistant synthetic tubes in these hoses are reinforced with four or six layers of spiraled, high-tensile steel wire over a layer of yarn braid.

Hose Size Specifications

Choose a hose with an inside diameter that is adequate to minimize pressure loss and to avoid hose damage caused by the heat generated by excessive fluid turbulence. A small hose increases pressure loss. An oversized hose adds unnecessary cost, weight, and bulk. The essential specifications of a hose include the inside diameter (ID), wall thickness (t), pressure ratings, and minimum bend radius.

Inside Diameter, Hoses

The size of a hose is specified by its inside diameter (ID). The ID must provide the proper volume of fluid for the specific application. The ID is specified in dash sizes or metric units. As given in a previous section, a dash number represents the size of a hose in terms of $1/16^{th}$ of an inch. For example, a hose with a $1/4$" (=$4/16$") inside diameter indicates that it has four numbers of $1/16$" segments. Therefore, its dash size would be a -4.

Pressure Rating, Hoses

The pressure rating of a hose is determined by its construction. The pressure rating is governed by its number of layers, materials, and method of construction. The more reinforcement layers a hose has, the more pressure it can stand. Burst pressure is attained when there is rupture of the hose or leakage from the end fitting. The rated pressure of a hose must be higher than the normal system pressure, and any pressure surges it will encounter.

Minimum Bend Radius, Hoses

The minimum bend radius is an essential consideration in the design and selection of a hose. Table 4.1 gives typical specifications of hoses.

Table 4.1 | Typical parameters of hoses in the English units

Dash Number	ID Inch	Work pressure bar psi	Min. Burst pressure bar psi	Min. bend radius mm inches
-2	1/8	3000	16000	
-3	3/16	3000	16000	
-4	1/4	3000	16000	1.5
-5	5/16	3000	16000	
-6	3/8	3000	16000	2.5
-8	1/2	3000	16000	2.9
-10	5/8	3000	16000	3.3
-12	3/4	3000	16000	4.0
-14	7/8	3000	16000	
-16	1	3000	16000	5.0
-20	1 1/4	3000	16000	12.0
-24	1½	3000	12000	14.0
-32	2	3000	12000	
-36	2 1/4	3000	12000	
-40	2½	3000	12000	
-48	3	3000	12000	
-56	3½	3000	12000	
-64	4	3000	12000	
-72	4½	3000	12000	

Type of Hose Motion

Typical hose motions are shown in Figure 4.3.

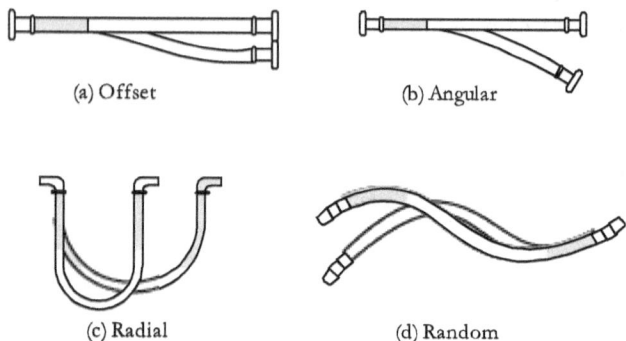

(a) Offset   (b) Angular

(c) Radial   (d) Random

Figure 4.3 | Types of hose motion

- When one end of the hose is moved in a plane perpendicular to its longitudinal axis with its ends remaining parallel is called offset motion.
- The angular motion of the hose occurs when one end of the hose is moved in a simple bend with its ends do not remain parallel
- The radial motion of the hose occurs when the hose is bent in a circular arc
- The random motion of the hose occurs in random planes

Standards of Hoses

ISO and SAE specify the standards for hoses to meet a set of dimensional and performance parameters.

- ISO 1436-1, wire braid reinforced
- ISO 4079-1, textile reinforced
- ISO 3949, thermoplastic textile reinforced
- ISO 3862-1, spiral wire reinforced

SAE J517 (US) standard has 100R numbers (SAE 100 R1 to SAE 100 R18) that define the specifications for the construction, dimension, pressure and temperature of hoses. Extracts of the specifications of SAE 100 R2 and R18 are presented below:

SAE 100 R2 specifications
The hose shall consist of an inner tube of oil-resistant synthetic rubber, steel wire reinforcement and an oil and weather resistant synthetic rubber cover. A ply or braid of suitable material may be used over the inner tube and /or over the wire reinforcement to anchor the synthetic rubber to the wire.

SAE 100 R18 specifications
The hose shall consist of a thermoplastic inner tube resistant to hydraulic fluids with suitable synthetic fiber reinforcement and hydraulic fluid and weather-resistant thermoplastic cover.

Table 4.2 gives the number and type of braids and pressure rating of hoses with various SAE numbers under SAE standards.

Table 4.2 | Braids and pressure ratings

SAE number	Braids	Pressure
100R1	1 Wire	500 – 2500 PSI
100R2	2 Wire	1200 – 3500 PSI
100R3	2 Cloth	375 – 1250 PSI
100R4	1 Wire Spiral	50 – 300 PSI
100R6	1 Cloth	300 – 600 PSI
100R12	4 wire	2500 – 5000 PSI

In Europe, the Committee for European Normalization (CEN / EN) standard is used to specify the flexible hose.
- EN 853 – Wire braided hose
- EN 854 - Fabric braided hose
- EN 855 – Thermoplastic textile braid hose
- EN 856 – Spiral wire hose
- EN 857 – Compact wire braided hose

Selection of Hoses

Selecting the right hose is the first step towards the safe operation and long service life of the associated hydraulic system. It all begins by choosing the right components, such as hoses, couplings, crimping equipment, and accessories. Hoses must meet the requirements of size, pressure, bend radius, and routing. It is also needed to know the equipment type, working and impulse pressures, fluid to be used, and bend radius.

Applications of Hoses

Hoses are used, when rigid or semi-rigid pipes or tubing cannot be used, as in applications with the movement of machine parts.

Advantages and Disadvantages, Hoses

Hoses offer many advantages. They are flexible and portable, and they absorb and dampen pressure surges and vibration. It is easy and faster to route hoses even around obstacles as compared to other types of conductors. They require no brazing or specialized bending.

However, mixing and matching couplings from one manufacturer with hoses from another manufacturer can lead to premature or catastrophic assembly failure. Hoses are susceptible to abuse, misapplication, and improper plumbing.

Hose Fittings

Hose fittings can be either permanent or reusable. Permanent hose fittings are installed on the hose by crimping and cannot be disassembled. Next, reusable hose fittings are screwed or clamped on the hose end.

The big difference between skive and the no-skive hose is in the thickness of the outer cover. The thicker cover requires a different fitting shell.

Fittings are made to Metric or SAE/JIC standards.

Quick Couplings (or Disconnects)

They are used for convenience as they can be installed and removed by hand and in situations where there is a need for the repeated connection and disconnection of the lines. A quick coupling has a male side and a female coupler. Quick couplings can be of the poppet type or flat face type.

In the poppet type, the male poppet (nipple) gets depressed when it engages with the coupler. This action opens the valve to allow the flow of hydraulic fluid. Poppet type quick couplers fall into ISO A or ISO B styles.

In the flat face couplers, both coupling sides have flat surfaces. A flat face male nipple will mate with a female flat face coupler. The back-end of flat face couplers can come with NPT, JIC, ORFS, or straight thread O-ring threads.

Based on the valving of the coupling, hydraulic couplings generally fall into one of the two groups: double shutoff, and straight through.

Double Shutoff couplings

They are extensively used when it is essential to minimize fluid loss upon disconnection. Both halves of the coupler, the body and the nipple, contain shutoff valves, as shown in Figure 4.4. These valves open automatically when the body and nipple are connected, and close automatically when the two halves are disconnected—keeping fluid loss to a minimum.

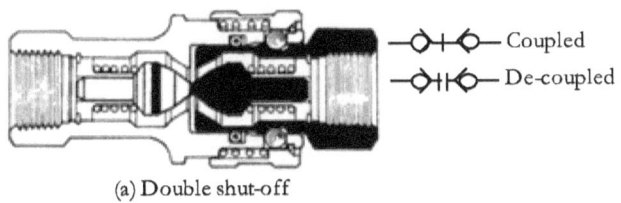

(a) Double shut-off

Figure 4.4 | A double shutoff coupling

Straight-Thru couplings

They have no valves in either half and are ideal for maximum flow application. Their smooth, open bore offers the lowest pressure drop of any quick disconnect coupling and allows them to be thoroughly cleaned.

Since there are no valves in either half, the fluid flow should be shut off before the coupling is disconnected.

Straight-through couplings are used where flow must be unrestricted. Figure 4.5 shows the cross-sectional view of a straight-through coupling.

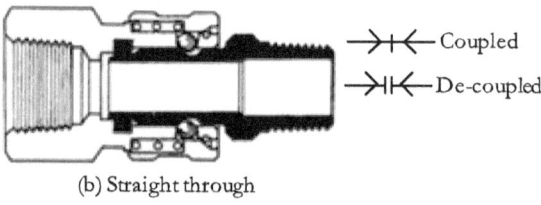

(b) Straight through

Figure 4.5 | A double straight-through coupling

Every manufacturer etches a part number into every coupler to help with its identification. It is better to consult the manufacturer's reference guide to ensure that quick disconnects (QDs) match up correctly.

Chapter 5 | Design of Hydraulic Piping Systems

A piping system for a hydraulic application is to be designed in conformity with many requirements of the application. In general, the suction pipe should be short and straight, and the return line should be large enough to limit the backpressure. When designing the piping system, the following factors have to be taken into account: (1) System pressure, (2) Operating temperature, (3) Duty cycle, (4) Shocks and vibration, (5) Material, (6) Connection technology, (7) Hoses and hose couplings, (8) Pipe supports, and (9) Standards.

Fluid Velocities – Suction Lines

The suction line is typically dimensioned so that the velocity does not exceed 1.2 m/s. The recommended fluid velocities for initial pipe sizing in suction lines are given in Table 5.1.

Table 5.1 | Fluid velocities in suction lines

Viscosity (cSt)	Maximum velocity (ft/s)
150	2
100	2.5
50	3.6
30	4

Fluid Velocities - Pressure Lines

The recommended fluid velocities for initial pipe sizing in pressure lines are given in Table 5.2.

Table 5.2 | Fluid velocities in pressure lines

Pressure line	For flow rate >2.6 gpm
900 – 1450 psi	13 – 14
1450 – 2300 psi	14 – 16
2300 – 3600 psi	16 – 18
3600 – 5800 psi	18 – 20

Fluid Velocities – Return Lines

Fluid velocities to be considered for the initial pipe sizing in return lines should be between 6.5 to 10 ft/s.

Dimensioning Based on Flow Velocity

When using the dimensioning method based on flow velocity, the inner diameter of the pipe can be determined by using the equation below, when a maximum flow rate and recommended flow velocity are known.

$$d = \sqrt{\frac{4 \times Qmax}{\Pi \times v}}$$

d   =Inner diameter of the pipe (ft)
Q_{max} =Maximum flow rate (ft³/s)
v   =Flow velocity (ft/s)

Example 5.1

Determine the inner pipe diameter to create a flow velocity of 15 ft/s with a flow rate of 0.0589 ft³/s.

Solution

Maximum Flow rate, Qmax   = 0.0589 ft³/s
Velocity, v   = 15 ft/s

$$\text{Pipe ID} = \sqrt{\frac{4 \times Qmax}{\Pi \times v}}$$

$$= \sqrt{\frac{4 \times 0.0589}{\Pi \times 15}}$$

$$= 0.0707 \text{ ft}$$
$$= 0.8484 \text{ in}$$

Dimensioning Based on Pressure Losses

When using the dimensioning method based on the pressure losses, the inner diameter of the pipe is selected so that the resulting pressure losses do not increase above a specified value. The total pressure loss is allowed to be 3 to 5% for systems in continuous use. The total pressure loss is allowed to be 7 to 10% for systems with intermittent duty cycle.

Flow Types

The amount of pressure losses also depends on the type of flow. When the flow is laminar, all the fluid particles move parallel to the pipe. When flow velocity increases, the flow will become turbulent, which means the direction of the individual fluid particles varies. The flow type can be found by determining the so-called Reynolds number (Re) and comparing it to critical Reynolds number value. The Reynolds number can be determined with the equation:

$$Re = \frac{v \cdot d \cdot \varrho}{\mu} = \frac{v \cdot d}{v}$$

Where,

Re $\quad$ = Reynolds number [-]
v $\quad$ = Flow velocity [ft/s]
d $\quad$ = Inner diameter of the pipe [in]
v (nu) $\quad$ = Kinematic viscosity [ft^2/s]
μ $\quad$ = Absolute viscosity [lb.s/ft^2]
ϱ $\quad$ = Density of the fluid [slug/ft^3]

The Reynolds number, with the kinematic viscosity in cSt, is given by:

$$Re = \frac{7740 \times v(ft/s) \times d(in)}{v \ (cSt)}$$

The flow is said to be laminar when Re < Re (critical).
The flow may be treated as turbulent when Re > Re (critical).

Re (critical) = 2000

Example 5.2

Determine the inner pipe diameter to create a flow velocity of 15 ft/s with a flow rate of 0.05886 ft³/s. A fluid with a density of 1.6 slug/ft³ and an absolute viscosity of 0.00668 lb.s/ft² is flowing through the pipe. Is the flow laminar or turbulent?

Solution

Q	$= 0.05886 \text{ ft}^3/\text{s}$
v	$= 15 \text{ ft/s}$
μ	$= 0.00668 \text{ lb.s/ft}^2$
ϱ	$= 1.6 \text{ slug/ft}^3$
A	$= Q/v = 0.05886 / 15 = 0.003924 \text{ ft}^2$
d	$= \sqrt{(4 \times A / \pi)} = 0.0707 \text{ ft} = 0.8484 \text{ in}$

$$\text{Reynolds number, Re} = \frac{v \cdot d \cdot \varrho}{\mu}$$

$$\text{Re} = \frac{15 \times 0.0707 \times 1.6}{0.00668}$$

$$\text{Re} = 254 \text{ (Laminar flow)}$$

Example 5.3

If fluid with a kinematic viscosity of 30 cSt is flowing at a velocity of 10 ft/s through a pipe of ¾-inch diameter, what would be the Reynolds number?

Solution

ν	$= 30 \text{ cSt}$
v	$= 10 \text{ ft/s}$
d	$= 0.75 \text{ in}$

$$\text{Re} = \frac{7740 \times v(\text{ft/s}) \times d(\text{in})}{\nu \text{ (cSt)}}$$

$$\text{Re} = \frac{7740 \times 10 \times 0.75}{30}$$

$$\text{Re} = 1935 \text{ (Laminar flow)}$$

Dimensioning Based on Pressure Losses

The overall pressure losses in the piping include frictional pressure losses arising in straight pipe sections, as well as individual pressure losses arising in bends and junctions.

Frictional Pressure Losses

The pressure losses in pipes and hoses can be estimated from the Darcy-Weisbach equation as given below:

$$\Delta Pa = \lambda \cdot \frac{l}{d} \cdot \frac{\varrho \, v^2}{2}$$

Δp_a	= Frictional pressure loss, [psi]
λ	= Frictional resistance factor [-]
l	= Length of the pipe [ft]
d	= pipe id [ft]
ϱ	= Hydraulic fluid density [Slug/ft^3]
v	= Flow velocity [ft/s]

Frictional Resistance Factor, λ

The friction factor is a function of Reynolds number and the surface roughness of round pipes and can be determined mathematically or by consulting the classic Moody Diagram. For laminar flow, the friction factor is independent of the surface roughness, and for turbulent flow, the friction factor is dependent both on the Reynolds number and the surface roughness.

For Laminar Flow

If the flow is laminar, λ depends only on the Reynolds number. The friction factor is given by:

$$\lambda = \frac{64}{Re}$$

Frictional Pressure Losses in Straight Pipe Sections

The pressure loss in a pipe for laminar flows can be determined by using the following equations in terms of the average velocity of flow or the fluid flow rate (more favorable), respectively:

$$\Delta Pa = \frac{32 \, \mu \, l \, v}{d^2}$$

$$\Delta Pa = \frac{128 \, \mu \, l \, Q}{\pi \, d^4}$$

Note: The pressure drop scales linearly with line length, therefore, long lines, and smaller diameter lines, could impact system efficiency.

Example 5.4

Calculate the frictional pressure loss for a 2.067" internal diameter pipe of length 100 ft through which fluid is flowing at a velocity of 3.8 ft/s. Kinematic viscosity = 44.4 cSt, and the density of the fluid 1.6 slug/ft³.

Solution

v	= 3.8 ft/s
d	= 2.067 inch = 0.17225 ft
ν	= 44.4 cSt
ϱ	= 1.6 slug/ft³
l	= 100 ft
Re	= 7740 . v . d / ν
	= 7740 x 3.8 x 2.067 /44.4 = 1369
λ	= 64/1369 = 0.0467
Δp_a	= λ (l/d) (ϱ.v²/2)
	= 0.0467x (100/0.17225) x (1.6 x 3.8²/2)
	= 0.0467 x 580.55 x 11.552
	= 313 lb/ft² = 313/144 psi = 2.17 psi

Frictional Losses in Turbulent Flow
The inside surface of a round pipe is given in Figure 5.1. The mean height of the roughness is designated as 'ε' and 'D' is the pipe inside diameter.

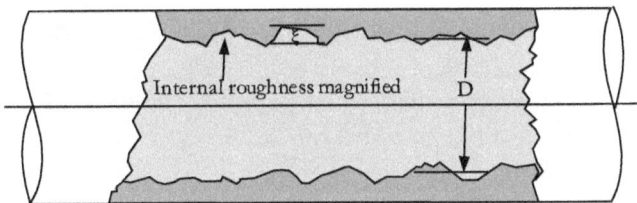

Figure 5.1 Relative roughness in a pipe

Relative roughness of the pipe's inside surface is defined as the mean roughness (ε) divided by the pipe inside diameter (D). That is,

Relative roughness $= \varepsilon/D$

Pipe roughness values depend on the pipe material as well as the method of its manufacture.

Typical values of absolute roughness:
Drawn tubing — 0.0015 mm
Cast iron — 0.26 mm
Riveted steel — 1.8 mm

Turbulent Flow, Smooth Pipes
In most fluid power systems, the pipes and hoses have smooth interiors, and the friction factor (λ) for smooth pipes can be calculated using the empirical formula given below.

$$\lambda = \frac{0.316}{Re^{0.25}}$$

The pressure loss (ΔPa) in a smooth pipe with the turbulent flow can now be calculated using the following formula:

$$\Delta Pa = 0.214 \, \frac{\mu^{0.25} \, 1 \, \varrho^{0.75} \, Q^{1.75}}{D^{4.75}}$$

Turbulent Flow, Rough Pipes

A close approximation of the friction factor for the turbulent flow through a pipe with rough interiors can be determined from the Swamee Jain equation given below:

$$\lambda = \frac{0.25}{[\log_{10}(\epsilon/3.7 \, d) + \left(5.74/_{Re^{0.25}}\right)]^2}$$

Example 5.5

Hydraulic 68 grade oil is flowing through a hydraulic line with inside diameter 0.0504 m at the rate of 0.0126 m³/s. Find the pressure drop for a 3.048 m length of hose. Assume the fluid density as 880 Kg/m³.

Solution

Q	= 0.445 ft³/s
d	= 0.17 ft
1	= 10 ft
ϱ	= 1.7 slug/ft³
ν	= 6.3 x 10⁻⁶ ft²/s
μ	= $\varrho\nu$ = 1.7 x 6.3x10⁻⁶ = 0.00001071 lb s/ft²
V	= 4Q/($\prod$d²)=4x0.445/(π x 0.17²) =19.6 ft/s

$$Re = v \, d/\nu = 19.6 \times 0.17/0.0000063 = 5289$$

$$\Delta Pa = 0.214 \, \frac{\mu^{0.25} \, 1 \, \varrho^{0.75} \, Q^{1.75}}{d^{4.75}}$$

$$\Delta Pa = 0.214 \, \frac{0.00001071^{0.25} \times 10 \times 1.7^{0.75} \times 0.445^{1.75}}{0.17^{4.75}}$$

$$= (0.214 \times 0.0572 \times 10 \times 1.4888 \times 0.2425) / 2.2 \times 10^{-4} = 200 \text{ psi}$$

Moody Diagram

The Moody Diagram, as given in Figure 5.2, plots the friction factor as a function of the Reynolds number and the relative pipe roughness. From the diagram, it can be inferred that for laminar flows, the friction factor depends only on the Reynolds number. For turbulent flows, the friction factor depends both on the Reynolds number and the relative roughness.

The following procedure can be followed to find the friction factor from the Moody Diagram:

- Find the value of Re
- Find the relative roughness ε
- Project a vertical line on the Re axis at the value of Re determined
- Find the curve corresponding to relative roughness
- Project horizontally to the f axis to obtain the friction factor

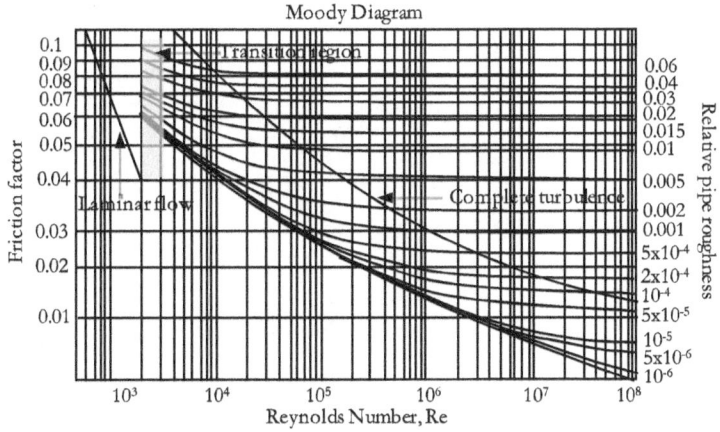

Figure 5.2 | Moody Diagram
Ref: Moody, L F (1944), 'Friction factors for pipe flow'

Individual Pressure Losses

The individual pressure losses occur in pipe bends, junctions and generally in pipe sections where the cross-sectional area or flow direction changes. The relation for the individual pressure losses is given by:

$$\Delta Pb = \zeta \frac{\varrho \cdot v^2}{2} = \zeta \frac{\varrho \cdot Q^2}{2\,A^2}$$

Where,

Δp_b	= individual pressure loss [Pa]
ζ	= individual resistance factor [-]
ϱ	= the fluid density [kg/m3]
v	= flow velocity [m/s]

Values of Loss Coefficient (ζ)

The value of the individual resistance factor ζ depends on the flow channel structure and dimensioning:

The loss coefficients (ζ):

- 90^0 elbow — 0.2
- 45^0 elbow — 0.15
- Tee fitting — 0.9
- Sharp-edged entrance — 0.5
- Rounded entrance — 0.05
- Sharp-edged exit — 1.0
- Rounded exit — 1.0

Individual pressure losses can also be found from the nomogram given in Figure 5.3

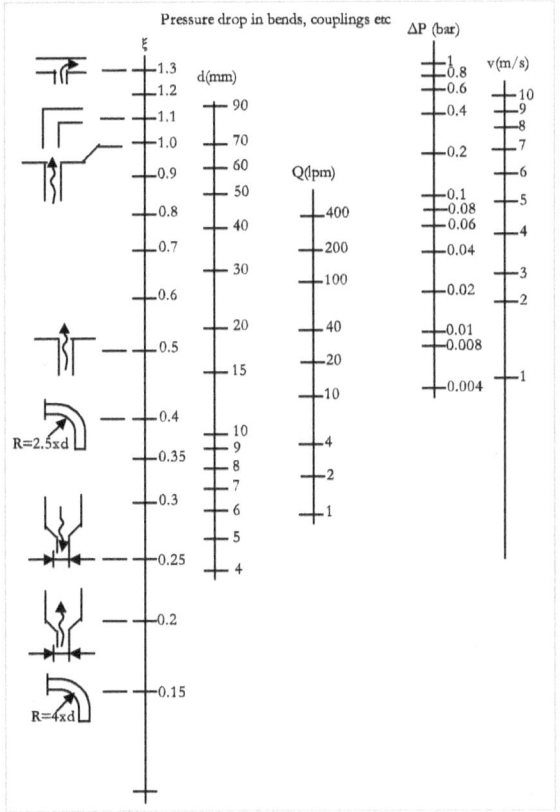

Figure 5.3 | Individual pressure losses

Total Pressure Losses

The total pressure losses in the piping are obtained by adding up the frictional pressure losses and the individual pressure losses.

$$\Delta Ptot = \Delta Pa + \Delta Pb$$

The total pressure loss is allowed to be 3 to 5% for systems in continuous use. The total pressure loss is allowed to be 7 to 10% for systems with intermittent duty cycle.

Chapter 6 | Installation, Routing, and Maintenance of Conductors

The proper routing, installation, and maintenance of fluid conductors and their fittings are as important as that of any other component in a hydraulic system. The service life of the fluid conductors is reduced by their installation stress, abrasion, tight bends, and exposure to higher pressure, temperature, and corrosion. One major problem with the fittings is the risk of leakage due to the loosening of the fittings as a result of shock and vibration in the system.

Installation of Hydraulic Conductors

It is essential to clean conductors and fittings before their installation. Remember to use a minimum number of fittings and connectors. The following bulleted lines give essential points for the proper installation of fluid conductors:

- Keep the length of conductors as small as possible to avoid tight bends,
- Route the fluid conductors optimally to minimize the pressure loss and leakage in the system, and reduce their abrasion, rubbing, kinking, and excessive flexing,
- Restrain, support, protect and guide fluid conductors using clamps at frequent intervals, to prevent chafing against one another and minimize the vibration of the conductor system,
- Avoid crossing two hose lines. However, when the crossing is unavoidable, join the two lines at the junction point,
- Use proper tools for preparing a conductor for connecting to other conductor or a component, and
- Conductors need to be flushed thoroughly with a suitable degreasing agent, immediately after their installation.

Hose Assembly Routing Tips

It is better to avoid sharp bends when routing hydraulic hoses. Using a bent or kinked hose causes severe backpressure in the associated system. Also, it may cause internal damage to the hose leading to its premature failure. The following bulleted lines list some tips for routing a hose assembly, especially about its length, minimum bend radii, and multi-plane bending:

- The hose length must be slightly longer than the actual distance between two linear connections to accommodate for the changes in the length with pressure changes
- The bend radius of the hose must be as large as possible to avoid the hose collapse or the flow restriction
- As far as possible, bend a hose in one plane only. This precaution is to avoid twisting of its wire reinforcement and improve its pressure capability. The multi-plane bending of a piece of the hose can often be avoided by rerouting the hose
- If the multi-plane bending cannot be avoided, install a clamp between the bends and provide enough hose length on both sides of the clamp for relieving the strain on hose's reinforcement wires
- Use clamps for securing the length of the hose in position and for keeping it from rubbing against adjacent surfaces
- Ensure that the hose is slack on both sides of the clamp for compensating for contraction and expansion
- The hose connected to a cylinder that undergoes pivoting motion must be of proper length to avoid its kinking or bending beyond its minimum bend radius
- Use appropriate swivel joints for the hose connection to reduce the bending force transmitted to the hose assembly by the relative motion between the associated machine elements
- The use of carriers keeps the hoses neatly nestled to prevent their rubbing against each other

Maintenance of Hydraulic Conductors

The following points give the maintenance activities of the hydraulic conductors and fittings, in a generalized manner. For exhaustive maintenance activities of the conductors, the reader may refer to the manufacturer's catalogues.

- Inspect the conductors and their routing for damages, defects, and displacement
- Examine the conductors and their joints for leakage, looseness, scratches, kinks, and burrs
- Tighten any loose fittings or nut connections with the correct amount of torque to minimize leakage and reduce contamination
- Repair or replace defective conductors or fittings. Ensure that any pipe replacement is of the same length, size, and wall thickness

Fluid Conductor Faults

Table 6.1 gives some faults in fluid conductors and the remedial actions.

Table 6.1 | Fluid conductor faults

Fault	Remedy
Weakened tubing or hose	-Use compatible fluid
Hose tube cracked	-Protect hose from excessive heat
Hose bursts	-Use the correct hose to withstand high-pressure surges -Alter environmental or operating conditions -Use correct length hose
Excessive pressure drop in hose	-Use hose of the correct size -Improve bore condition
Faulty fittings	-Replace faulty ones
Leakage through fittings	-Replace damaged seal -Correct damaged threads -Use clamps to prevent loosening of fittings

7 | Objective Type Questions

1. The schedule number is associated with:
   a) The wall thickness of a pipe
   b) Viscosity index
   c) The acidity level of a hydraulic fluid
   d) The hardness of a seal material

2. Mark the least expensive hydraulic fluid conductor system?
   a) Pipes
   b) Tubing
   c) Hoses
   d) Cannot be distinguished

3. The dash number is used to specify:
   a) Inside diameter (ID) of a hose
   b) Wall thickness of the tubing
   c) The elasticity of seal materials
   d) Contamination concentration level

4. Mark the incorrect statement
   a) Hoop stress of a given length of pipe is the circumferential stress acting on the wall of the pipe under the operating pressure.
   b) Bending hose/tubing to a smaller radius than its rated minimum bend radius may result in its premature failure.
   c) Flared or flareless-type fittings can be used for tubing end-connections.
   d) Hose length must be precisely equal to the actual distance between their end connections.

8 | Review Questions

1. What is the function of fluid conductors in hydraulic systems?
2. List three primary types of fluid conductors used in hydraulic systems. Briefly explain about their degree of flexibility.
3. What are the basic requirements for the satisfactory function of the fluid conductor system in a hydraulic circuit?
4. State the reason for the probable energy loss in the fluid conductors used in hydraulic systems.
5. What are the reasons for the fluid leakages in a hydraulic distribution system?
6. What are the ways to minimize fluid leakage in a hydraulic fluid distribution system?
7. What is the definition of 'schedule number' when referring to piping and fittings in hydraulic systems?
8. What variables determine the wall thickness and the factor of safety of a fluid conductor?
9. Why should fluid conductors have greater strength than the system working pressure requires?
10. Briefly, explain the terms of fluid conductors: (1) Bend radius, (2) Tensile stress, (3) Burst pressure, and (4) Working pressure.
11. What factors determine the pressure rating of a fluid conductor?
12. List out the procedure to calculate the size of a fluid conductor for a hydraulic system.
13. Explain the purpose of hydraulic pipes, briefly.
14. State two disadvantages of using pipes in hydraulic systems
15. How the pipes used as fluid conductors in hydraulic systems, are specified?
16. Explain how the wall thickness of the pipe used as a fluid conductor in a hydraulic system, specified.
17. What is meant by the schedule number of a standard pipe, as used in hydraulic systems?
18. How is the pipe size classified in hydraulic systems?
19. Describe the methods of coupling pipes in hydraulic systems.

20. What are the functions of pipe threads, as used in hydraulic systems?
21. State some common materials used in the manufacturing of hydraulic pipes.
22. State the major disadvantages of steel pipes, as used in hydraulic systems.
23. What are the two types of thread configurations used in the piping systems for hydraulic systems? Differentiate them.
24. Name two common types of pipe joints used in hydraulic systems.
25. What is the disadvantage of threaded fittings for hydraulic systems?
26. Briefly explain the use of tubing in hydraulic systems.
27. Why is steel tubing more commonly used than steel pipe in hydraulic systems?
28. State the common materials used for manufacturing hydraulic tubing.
29. How do you specify hydraulic tubing?
30. What factors determine the tubing size?
31. Distinguish between the thin-walled and the thick-walled hydraulic conductors.
32. Mention one advantage and one disadvantage of hydraulic tubing.
33. Mention two advantages of the hydraulic tubing over the pipes
34. Describe any two methods of coupling the tubing in hydraulic systems.
35. Describe different types of tube fittings, as used in hydraulic systems.
36. Briefly explain the correct methods for bending and flaring hydraulic tubing.
37. What is a flare fitting, as used in hydraulic systems? Why is flaring needed, and how is it done?
38. What is the difference between the flared fitting and the compression fitting, as used in hydraulic systems?
39. List the parts of a flared tubing fitting assembly, as used in hydraulic systems.

40. List the parts of a flare-less tubing fitting assembly, as used in a hydraulic system.
41. Briefly explain the use of hoses in hydraulic conductor systems.
42. Mention three essential elements of a flexible hose, as used in hydraulic systems.
43. Describe the basic constructional features of flexible hoses, as used in hydraulic systems?
44. What is the purpose of providing a protective outer layer for a hydraulic hose?
45. Under what conditions flexible hoses are used in hydraulic systems?
46. How the hoses used as fluid conductors in hydraulic systems, are specified?
47. What does the dash number of a hydraulic hose refer to?
48. What determines the pressure rating of a hydraulic hose?
49. Explain how the pressure rating of hydraulic hoses is increased.
50. What are the advantages of hydraulic hoses?
51. Mention three factors that are to be considered while selecting hydraulic hoses?
52. Briefly explain five essential parameters to be considered while selecting a hydraulic hose.
53. Give a brief note on (1) Hose motion in hydraulic systems and (2) The applications of hydraulic hoses.
54. Explain the purpose of the quick disconnect coupling, as used in hydraulic systems.

9 | Numerical Problems

1. Calculate the minimum inside diameter of the suction pipe in a hydraulic system to handle the flow rate of 10.6 gpm with an average fluid velocity not exceeding 1.6 ft/s. [Ans: 1.65 in]

2. Determine the size of the pressure line of a hydraulic system with a 10 gpm positive-displacement pump. The recommended flow velocity through the pressure line is 16.4 ft/s. [Ans: 0.5 in]

3. Determine the size of the return line of a hydraulic system with a 16.25 gpm positive-displacement pump. The recommended return flow velocity is 8.2 ft/s. [Ans: 0.9 in]

4. A hydraulic system is to permit the flow rate of 10.6 gpm with an average fluid velocity not exceeding 13 ft/s. Calculate the minimum inside diameter of the pressure conductor in the system. [Ans: 0.575 in]

5. A hydraulic system is to permit the flow rate of 10.6 gpm with an average fluid velocity not exceeding 6.5 ft/s. Calculate the minimum inside diameter of a return-line conductor. [Ans: 0.814 in]

6. Find the schedule number of a steel pipe for a hydraulic system at the estimated working pressure of 1625 psi. The allowable stress is 58000 psi.[Ans: 30]

7. Calculate the burst pressure of a seamless cold-drawn steel tubing of outside diameter 0.98 in and wall thickness 0.098. The tubing has a tensile strength of 58000 psi. [Ans: 12889 psi]

8. Determine the safe working pressure for a steel tube with a burst pressure of 13000 psi, assuming a safety factor of 8. [Ans: 1625 psi]

9. A hydraulic pipe has an outer diameter of half an inch and schedule 40, and another pipe has the same outer diameter but of schedule 80. State the difference between these two pipes, explicitly.

10 | References

1. Anthony Esposito, Fluid Power with Applications, 6th Edition, Prentice-Hall of India, 2006
2. Article on: 'About Hydraulic Hose', GlobalSpec Inc., Jordan Rd, Troy, NY, USA
3. Article on: 'Hose and tubing assemblies, Hydraulic hose, Hose installation', Hydraulics & Pneumatics Magazine, The Penton Media Building, Cleveland, OH USA
4. Article on: 'Hydraulic system tubing – Lifelines to power and motion control', by Terry Karl and Mark Morrow
5. Catalogue on 'High-Pressure Stainless Steel Hoses' PARKER / PAGE International Hose, Texas, www.pageintl.com
6. Catalogue on 'Hose and Flexible Tubing', Document No. R8 MS-01-180, Swagelok Company, U.S.A., AGS October 2013
7. Catalogue on: 'Hose, Fittings, and adapters catalogue 2010' by Alfagomma Hydraulic spa, Vimercate, MI, Italy, http://www.alfagomma.com/
8. Paper on: 'Dash Number Chart', Jones Enterprise, LaPorte, Indiana, USA
9. Document on: 'Flexible metal hoses', HAM-LET Advanced Control Technology, info@ham-let.com
10. Document on: 'How We Bend Steel Tubing and Steel Beams', Paramount Roll and Forming Inc, Los Angeles, California
11. Document on: 'Hydraulic Hose Installation', Airline Hydraulics Corporation, Bensalem, PA, USA
12. Document on: 'PROPER INSTALLATION & HOSE ROUTING', Good Year, www.hydraulics.goodyear.com
13. Documents on: 'Introduction to Hydraulics', 'Hydraulic Hose Life Made Simple', GATES Professional Development series, The Gates Rubber Company, Colorado, USA
14. William D. Wolansky et al., Fundamentals of Fluid power, Houghton Mifflin Company, Boston, 1977

Fluid Power Educational Series Books

1. Pneumatic Systems and Circuits -Basic Level (In the SI Units)
2. Industrial Pneumatics -Basic Level (In the English Units)
3. Pneumatic Systems and Circuits -Advanced Level
4. Electro-Pneumatics and Automation
5. Design of Pneumatic Systems (In the SI Units)
6. Design Concepts in Pneumatic Systems (In the English Units)
7. Maintenance, Troubleshooting, and Safety in Pneumatic Systems
8. Industrial Hydraulic Systems and Circuits -Basic Level (In the SI Units)
9. Industrial Hydraulics -Basic Level (In the English Units)
10. Hydraulic Fluids
11. Hydraulic Filters: Construction, Installation Locations, and Specifications
12. Hydraulic Power Packs (In the SI Units)
13. Power Packs in Hydraulic Systems (In the English Units)
14. Hydraulic Cylinders (In the SI Units)
15. Hydraulic Linear Actuators (In the English Units)
16. Hydraulic Motors (In the SI Units)
17. Hydraulic Rotary Actuators (In the English Units)
18. Hydraulic Accumulators and Circuits (In the SI Units)
19. Accumulators in Hydraulic Systems (In the English Units)
20. Hydraulic Pipes, Tubes, and Hoses (In the SI Units)
21. Pipes, Tubes, and Hoses in Hydraulic Systems (In the English Units)
22. Design of Industrial Hydraulic Systems (In the SI Units)
23. Design Concepts in Industrial Hydraulic Systems (In the English Units)
24. Maintenance, Troubleshooting, and Safety in Hydraulic Systems
25. Hydrostatic Transmissions (HSTs) (In the SI Units)
26. Concepts of Hydrostatic Transmissions (In the English Units)
27. Load Sensing Hydraulic Systems (In the SI Units)
28. Concepts of Load Sensing Hydraulic Systems (In the English Units)
29. Electro-hydraulic Proportional Valves
30. Electro-hydraulic Servo Valves
31. Cartridge Valves
32. Electro-hydraulic Systems and Relay Circuits

For more details, please visit: **htpps://jojibooks.com**

About the Author

Joji Parambath is a trainer in the field of Pneumatics, Hydraulics, and PLC, for over 25 years. During his career, he has trained numerous professionals from the industries as well as faculty members and students of engineering institutions.

At present, he is the key trainer at Fluidsys Training Centre, Bangalore, India, (https://fluidsys.org) which is providing training in the field of Pneumatics and Hydraulics. He has already written two books on Pneumatics and Hydraulics. The publication of the present series of 32 books is intended to restructure and update the existing books.

The author wishes to thank all trainees for their lively interaction and many useful suggestions during the training programmes that prompted the author to write the present series of books. You may send your feedback to joji.p@hotmail.com

10th June 2020